TABLE OF CONTENTS

3-4	**INSTRUCTIONS**
5-6	**STEGOSAURUS** 
7-8	**TRICERATOPS** 
9-10	**ALLOSAURUS** 
11-12	**DIPLODOCUS** 
13-14	**TYRANNOSAURUS REX** 
15-16	**SPINOSAURUS** 
17-18	**VELOCIRAPTOR** 
19-20	**QUETZALCOATLUS** 
21-22	**PARASAUROLOPHUS** 
23-24	**COMPSOGNATHUS** 
25-28	**DIGITAL TOYS**

HOW IT WORKS

What are Popar® books?

Popar® books use Augmented Reality (AR) technology to create a immersive reading experience that will allow the user to see their books come alive with incredible virtually "real" 3D objects, animations, and interactions that will pop off the page. Popar® books are designed to change the way we interact and experience stories, adventures, and learning.

What is Augmented Reality?

Our Augmented Reality (AR) is a ground breaking concept that uses a mobile/smart device and special patterns to make amazing and engaging 3D objects, animations, and interactions appear in the real world that maintain interest in all Popar® book series.

What do I need to see AR?

-Mobile device that meets the minimum system requirements:

Apple® - iPhone® 4 or higher, iPad® 2 or higher

Android™ - Smart phone or tablet w/ camera & Google Play™

-Included Popar® book software

-Popar® book

1 FIND

Use Mobile/Smart Device
Minimum requirements: Apple® iPhone® 4 or higher, Apple® iPad® 2 or higher, or Android™ smart phone or tablet w/ camera & Google Play™ to activate the Augmented Reality and other technology.

2 DOWNLOAD

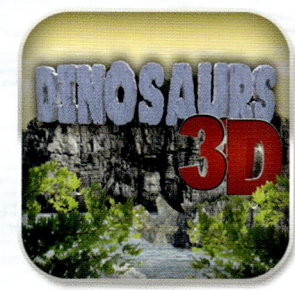

Download the FREE Popar® App
Apple® users use the App Store and Android™ users use Google Play™. Search for Popar® Dinosaurs 3D.

3 EXPLORE

Look For Special Symbols
These symbols will show you where there are Augmented Reality (AR) interactions, videos, read-alongs and games!

4 PLAY

Focus Your Mobile/Smart Device
Focus your mobile/smart device **10 INCHES** away from the special Augmented Reality (AR) symbols and watch the magic happen.

FEATURES   FIND THIS

All Popar® books are packed full of fun, exciting, educational, and interactive features. As you learn and play with each action packed page, make sure you are keeping your eye out for the above special AR symbols. These symbols will show you where there is immersive Augmented Reality content waiting to be explored. Make sure to read the below instructions so you can learn more about the amazing 3D features that await exploration.

INTERACT

When you find one of the special AR symbols, you can interact with the 3D objects or animations simply with hand gestures on your mobile/smart device.

Touch
You can change an animation or object simply by touching the screen.

Swipe
You can also spin each object or animation by using your finger to swipe the screen.

Pinch
Enlarge or shrink the size of the 3D object or animation by pinching or expanding two fingers on your screen.

READ-ALONG

Every Popar® book has an educational read-along feature. Simply locate one of the special AR symbols on every page, and touch the AR enhanced area on the screen in order for the book to read to you. This feature is great for auditory learners. Learning and playing has never been so easy.

VIDEO

Popar® books are also filled with amazing educational videos that appeal to visual learners. When you find that special AR symbol, be prepared because you are in the front seat of the movie theater as a screen will appear right on the pages of the book giving you an up close and personal video experience.

4

THE STEGOSAURUS WAS A PLANT EATING ANIMAL (HERBIVORE). SPIKES ATTACHED TO THE END OF ITS TAIL HELPED WITH FENDING OFF OTHER PREDATORS. BECAUSE OF THE LARGE SIZE OF THE ANIMAL AND ITS STOCKY LEGS, THE STEGOSAURUS HAD A VERY SMALL STRIDE AND COULD NOT RUN VERY QUICKLY. CONSIDERING THE LARGE SIZE OF THIS DINOSAUR, THE STEGOSAURUS WOULD HAVE HAD TO CONSTANTLY EAT LOTS OF VEGETATION TO GET ITS DAILY FILLING. STEGOSAURUSES PROBABLY TRAVELED IN PACKS AS ANOTHER PROTECTION BEHAVIOR.

THEY USUALLY WEIGHED AROUND 6,000 POUNDS. LARGE, SHIELD-SHAPED PLATES COMING OUT OF ITS BACK MADE THEM VERY IDENTIFIABLE. THE PLATES STAGGERED FROM THE LEFT SIDE OF THE DINOSAUR'S BODY TO THE RIGHT SIDE AND CONTAINED BLOOD VESSELS.

STEGOSAURUS

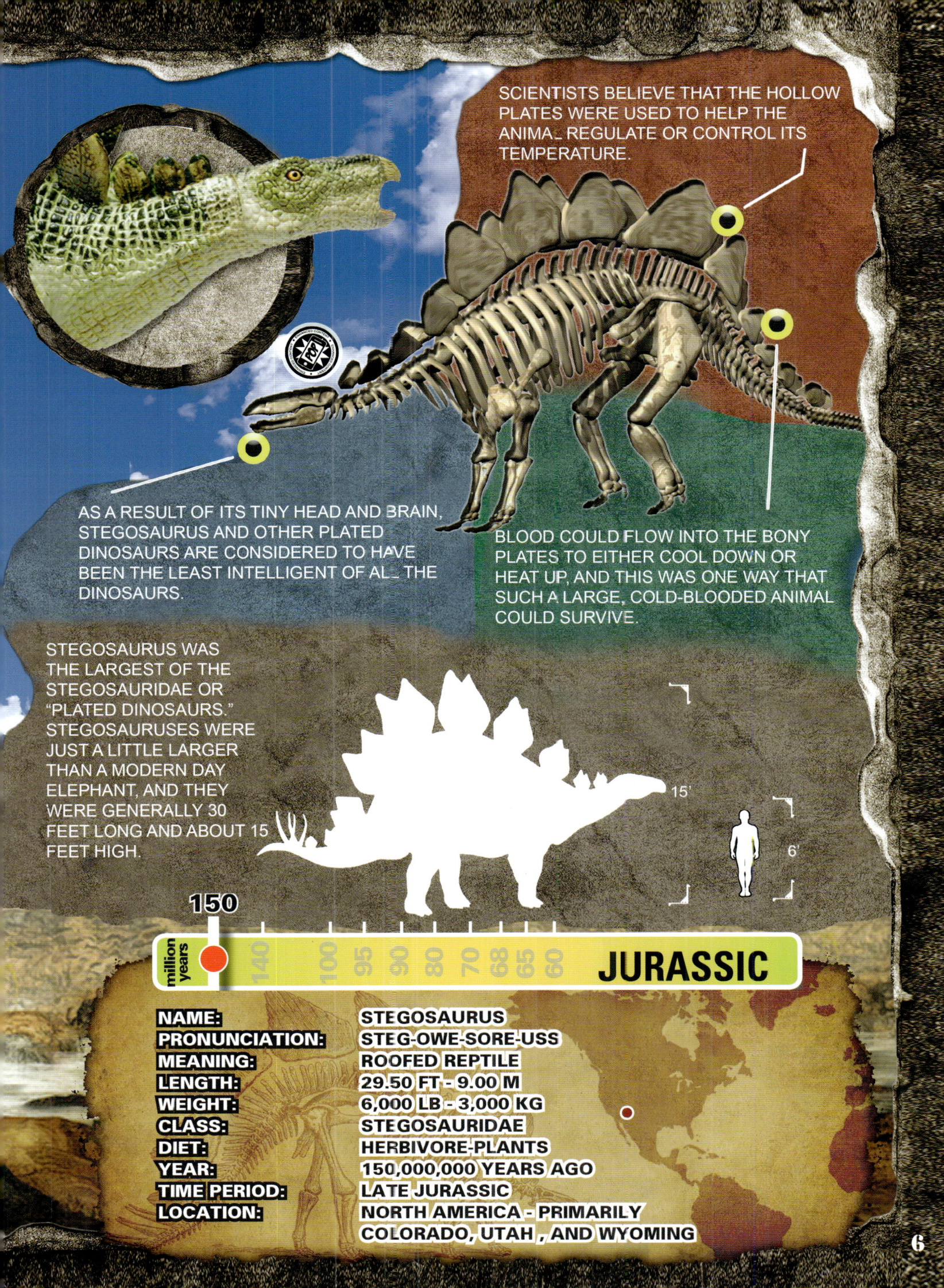

SCIENTISTS BELIEVE THAT THE HOLLOW PLATES WERE USED TO HELP THE ANIMAL REGULATE OR CONTROL ITS TEMPERATURE.

AS A RESULT OF ITS TINY HEAD AND BRAIN, STEGOSAURUS AND OTHER PLATED DINOSAURS ARE CONSIDERED TO HAVE BEEN THE LEAST INTELLIGENT OF ALL THE DINOSAURS.

BLOOD COULD FLOW INTO THE BONY PLATES TO EITHER COOL DOWN OR HEAT UP, AND THIS WAS ONE WAY THAT SUCH A LARGE, COLD-BLOODED ANIMAL COULD SURVIVE.

STEGOSAURUS WAS THE LARGEST OF THE STEGOSAURIDAE OR "PLATED DINOSAURS." STEGOSAURUSES WERE JUST A LITTLE LARGER THAN A MODERN DAY ELEPHANT, AND THEY WERE GENERALLY 30 FEET LONG AND ABOUT 15 FEET HIGH.

JURASSIC

NAME:	STEGOSAURUS
PRONUNCIATION:	STEG-OWE-SORE-USS
MEANING:	ROOFED REPTILE
LENGTH:	29.50 FT - 9.00 M
WEIGHT:	6,000 LB - 3,000 KG
CLASS:	STEGOSAURIDAE
DIET:	HERBIVORE-PLANTS
YEAR:	150,000,000 YEARS AGO
TIME PERIOD:	LATE JURASSIC
LOCATION:	NORTH AMERICA - PRIMARILY COLORADO, UTAH, AND WYOMING

THE TRICERATOPS TRAVELED IN HERDS FROM LOCATION TO LOCATION DEPENDING ON FOOD AVAILABILITY. THE HORNS AND FRILLS OF A TRICERATOPS HAD MANY FUNCTIONS OTHER THAN DEFENSE INCLUDING COURTSHIP RITUALS, PACK DOMINANCE, STATUS SYMBOL, AND INCREASED BODY AREA. THEY ALSO USED THEIR HORNS DEFENSIVELY AGAINST LARGER ANIMALS.

OTHER DINOSAURS LIVING IN THAT TIME PERIOD INCLUDED ANKYLOSAURUS, CORYTHOSAURUS, DRYPTOSAURUS, AND THE POPULAR TYRANNOSAURUS REX. MOST TRICERATOPS HAVE BEEN FOUND ALONG WESTERN NORTH AMERICA.

AS A RESULT OF ITS FOUR STUMPY FEET, THE TRICERATOPS COULD ONLY RUN ABOUT 15 MILES PER HOUR.

THE TRICERATOPS, CLASSIFIED AS AN HERBIVORE (PLANT EATING), USED ITS STURDY BEAK TO EAT THROUGH STRONGER PLANTS AND LARGER SHRUBS.

CRETACEOUS

- **NAME:** TRICERATOPS
- **PRONUNCIATION:** TRY-SERRA-TOPS
- **MEANING:** THREE HORNED HEAD
- **LENGTH:** 29.50 FT - 9.00 M
- **WEIGHT:** 24,000 LB - 11,000 KG
- **CLASS:** CERATOPSIA
- **DIET:** HERBIVORE - PLANTS
- **YEAR:** 65,000,000 YEARS AGO
- **TIME PERIOD:** LATE CRETACEOUS
- **LOCATION:** NORTH AMERICA - PRIMARILY COLORADO AND WYOMING

TRICERATOPS

THE NAME TRICERATOPS LITERALLY MEANS "THREE HORNED FACE." THEY HAD TWO LARGE HORNS ON THE TOP OF THEIR SKULL AND A FINAL HORN ON THE TIP OF ITS BEAK SIMILAR TO A RHINOCEROS.

ITS SKULL ALONE COULD BE UP TO 10 FEET LONG.

LIKE OTHER HORNED DINOSAURS, TRICERATOPS HAD A LARGE RIDGE OR FRILL THAT RAN AROUND THE BACK OF ITS SKULL. THIS FRILL WAS MUCH LIKE A SHIELD OF BONE THAT WAS ABOVE ITS HEAD.

THE TRICERATOPS WAS ABOUT 10 FEET HIGH AND 30 FEET LONG. THEY COULD WEIGH UP TO 24,000 POUNDS!

ALLOSAURUS IS ONE OF THE LARGEST AND MOST FEARSOME PREDATORS TO HAVE EVER LIVED ON LAND. HUNDREDS OF SKELETONS HAVE BEEN FOUND THROUGHOUT NORTH AMERICA, AFRICA, AND AUSTRALIA, UNLIKE ANOTHER COMMON PREDATOR KNOWN AS THE TYRANNOSAURUS REX WHO HAS LESS THAN TWENTY FULL SETS OF BONES DISCOVERED TO DATE.

ITS SKULL WAS RELATIVELY SMALL FOR A PREDATOR OF ITS SIZE; HOWEVER, THIS MAY HAVE ALLOWED THIS DINOSAUR TO SPRINT MUCH FASTER AFTER ITS PREY.

ITS EYES WERE SPACED WIDE APART, SO ITS BINOCULAR VISION AND DEPTH PERCEPTION CANNOT HAVE BEEN PARTICULARLY KEEN.

ALLOSAURUS

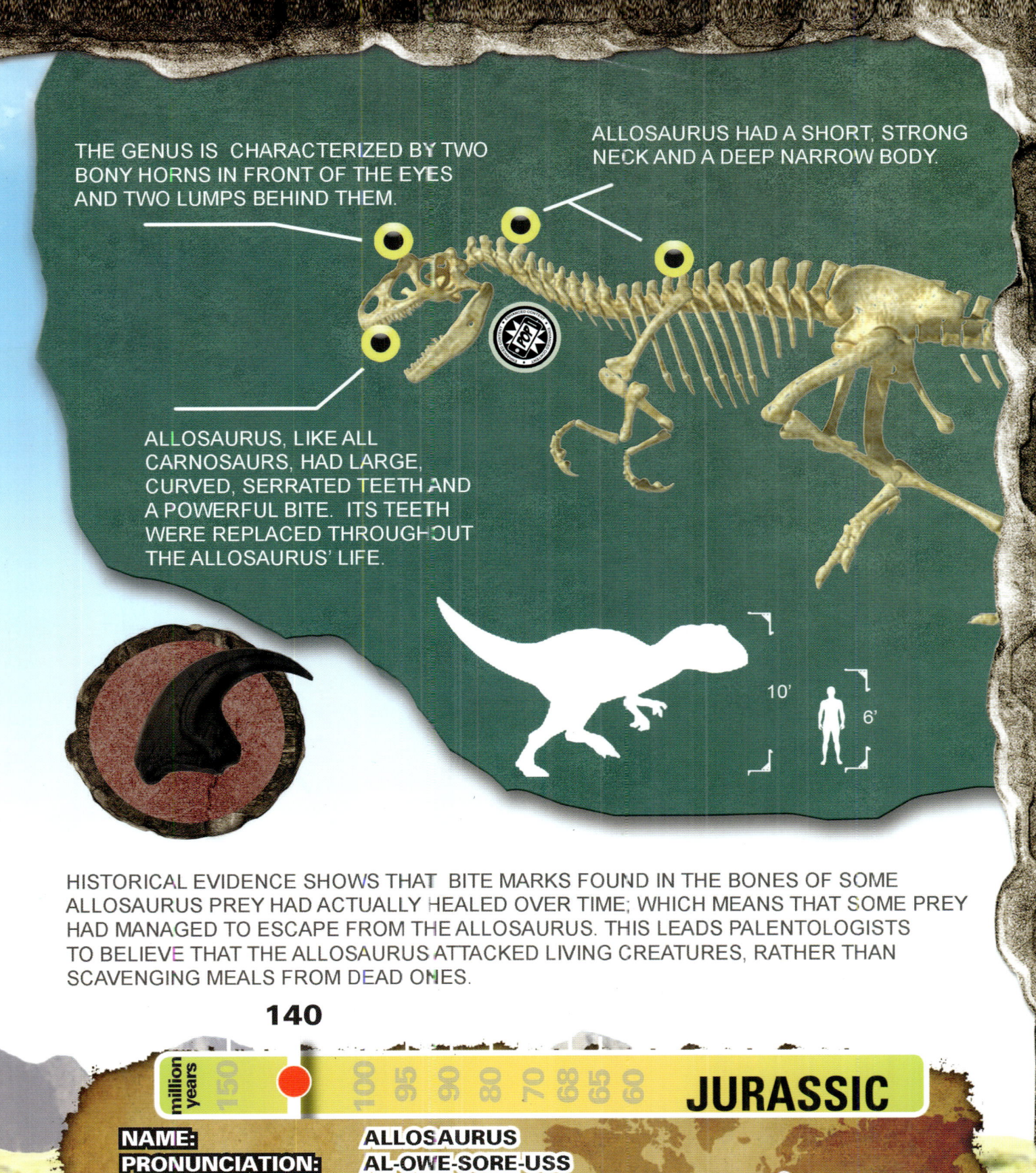

THE GENUS IS CHARACTERIZED BY TWO BONY HORNS IN FRONT OF THE EYES AND TWO LUMPS BEHIND THEM.

ALLOSAURUS HAD A SHORT, STRONG NECK AND A DEEP NARROW BODY.

ALLOSAURUS, LIKE ALL CARNOSAURS, HAD LARGE, CURVED, SERRATED TEETH AND A POWERFUL BITE. ITS TEETH WERE REPLACED THROUGHOUT THE ALLOSAURUS' LIFE.

HISTORICAL EVIDENCE SHOWS THAT BITE MARKS FOUND IN THE BONES OF SOME ALLOSAURUS PREY HAD ACTUALLY HEALED OVER TIME; WHICH MEANS THAT SOME PREY HAD MANAGED TO ESCAPE FROM THE ALLOSAURUS. THIS LEADS PALENTOLOGISTS TO BELIEVE THAT THE ALLOSAURUS ATTACKED LIVING CREATURES, RATHER THAN SCAVENGING MEALS FROM DEAD ONES.

140

JURASSIC

NAME:	ALLOSAURUS
PRONUNCIATION:	AL-OWE-SORE-USS
MEANING:	DIFFERENT REPTILE
LENGTH:	39.00 FT - 12.00 M
WEIGHT:	4,000 LB - 2,000 KG
CLASS:	THEROPODA
DIET:	CARNIVORE - LARGE ANIMALS
YEAR:	140,000,000 YEARS AGO
TIME PERIOD:	LATE JURASSIC
LOCATION:	AFRICA, AUSTRALIA, EUROPE, AND NORTH AMERICA - PRIMARILY COLORADO AND UTAH

DIPLODOCUS

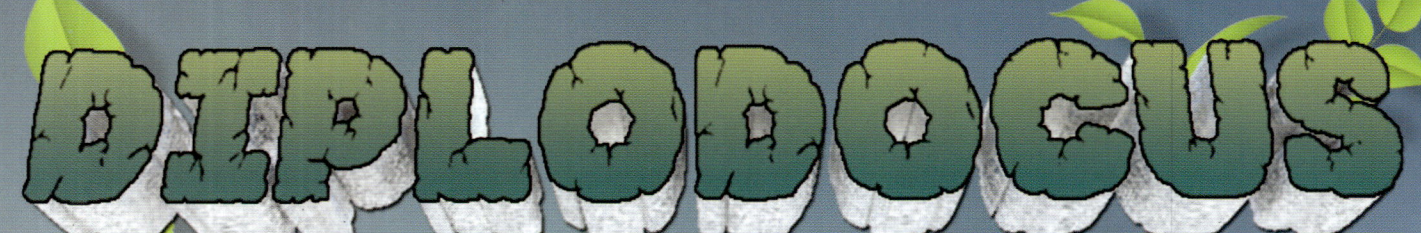

DIPLODOCUS IS THOUGHT TO BE THE LONGEST KNOWN DINOSAUR. THEY LIVED ABOUT 150 TO 154 MILLION YEARS AGO IN THE LATE JURASSIC PERIOD. THIS WAS ABOUT THE SAME TIME AS THE STEGOSAURUS AND THE ALLOSAURUS. THEIR SKELETONS HAVE BEEN DISCOVERED THROUGHOUT NORTH AMERICA, PRIMARILY IN COLORADO, UTAH, MONTANA AND WYOMING.

THIS LATE JURASSIC SAUROPOD THRIVED IN A TIME WHEN THE EARTH'S WARM TEMPERATURES ENCOURAGED RAPID PLANT GROWTH. DIPLODOCUS NEEDED A LOT OF FOOD TO KEEP IT ALIVE. DIPLODOCUS WERE PLANT EATERS (HERBIVORES) WHICH GRAZED IN LARGE HERDS TOGETHER USING THEIR RAKE-LIKE TEETH TO STRIP AN AREA OF VEGETATION VERY QUICKLY.

154 million years — JURASSIC

NAME: DIPLODOCUS
PRONUNCIATION: DIP-LOD-IC-USS
MEANING: DOUBLE BEAM
LENGTH: 90.00 FT - 27.00 M
WEIGHT: 24,000 LB - 11,000 KG
CLASS: SAUROPODA
DIET: HERBIVORE - PLANTS
YEAR: 154,000,000 YEARS AGO
TIME PERIOD: LATE JURASSIC
LOCATION: NORTH AMERICA - PRIMARILY COLORADO, UTAH, MONTANA AND WYOMING

THE TYRANNOSAURUS GENERALLY LIVED IN LARGER, FOREST CLIMATES BECAUSE THIS IS WHERE ITS PREY (PLANT EATING DINOSAURS) COULD BE FOUND MUCH EASIER. FOR ALMOST 90 YEARS, T-REX WAS THE LARGEST KNOWN PREDATOR TO HAVE EVER WALKED ON THE EARTH. THIS NOTORIETY WAS LOST IN THE 1990'S WHEN A LARGER PREDATOR, GIGANOTOSAURUS, WAS DISCOVERED IN ARGENTINA. FOSSILS OF THE T-REX HAVE BEEN FOUND IN MONGOLIA AND WESTERN NORTH AMERICA. THE LARGEST, MOST COMPLETE FOSSIL OF THE TYRANNOSAURUS WAS FOUND IN SOUTH DAKOTA, UNITED STATES. TYRANNOSAURUS REX WAS ONE OF THE LAST DINOSAURS TO DIE OUT.

BEING A LATE THEROPOD, IT HAD JUST TWO DIGITS OR FINGERS IN ITS UPPER FOREARMS. THE EARLY THEROPODS HAD FOUR DIGITS. WITH EVOLUTION, THEY SLOWLY LOST FINGERS.

THE TYRANNOSAURUS REX (T-REX) WAS A GIANT ANIMAL AT ABOUT 15 FEET TALL AND 40 FEET LONG. THAT IS A LITTLE BIT LONGER THAN A TYPICAL SCHOOL BUS.

IN ORDER TO REDUCE WEIGHT, THEIR HEAD WAS FULL OF HOLES, AN ADAPTATION SEEN IN MANY DINOSAURS.

TYRANNOSAURUS REX ATE WITH 50 VERY LARGE (UP TO NINE INCHES) TEETH.

THE T-REX'S HEAD ALONE WAS FOUR TO FIVE FEET IN LENGTH.

TYRANNOSAURUS REX

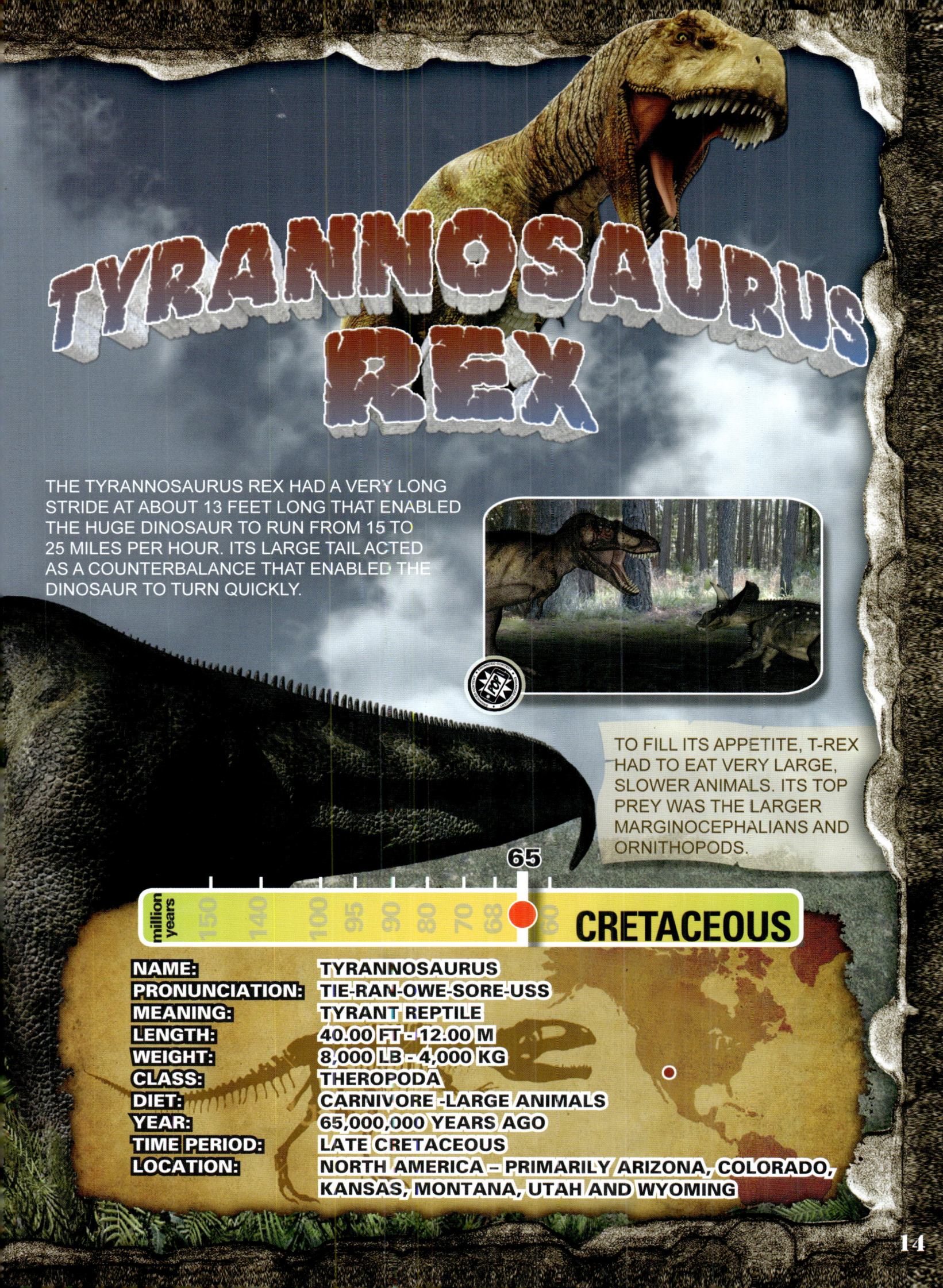

THE TYRANNOSAURUS REX HAD A VERY LONG STRIDE AT ABOUT 13 FEET LONG THAT ENABLED THE HUGE DINOSAUR TO RUN FROM 15 TO 25 MILES PER HOUR. ITS LARGE TAIL ACTED AS A COUNTERBALANCE THAT ENABLED THE DINOSAUR TO TURN QUICKLY.

TO FILL ITS APPETITE, T-REX HAD TO EAT VERY LARGE, SLOWER ANIMALS. ITS TOP PREY WAS THE LARGER MARGINOCEPHALIANS AND ORNITHOPODS.

CRETACEOUS

NAME:	TYRANNOSAURUS
PRONUNCIATION:	TIE-RAN-OWE-SORE-USS
MEANING:	TYRANT REPTILE
LENGTH:	40.00 FT - 12.00 M
WEIGHT:	8,000 LB - 4,000 KG
CLASS:	THEROPODA
DIET:	CARNIVORE -LARGE ANIMALS
YEAR:	65,000,000 YEARS AGO
TIME PERIOD:	LATE CRETACEOUS
LOCATION:	NORTH AMERICA – PRIMARILY ARIZONA, COLORADO, KANSAS, MONTANA, UTAH AND WYOMING

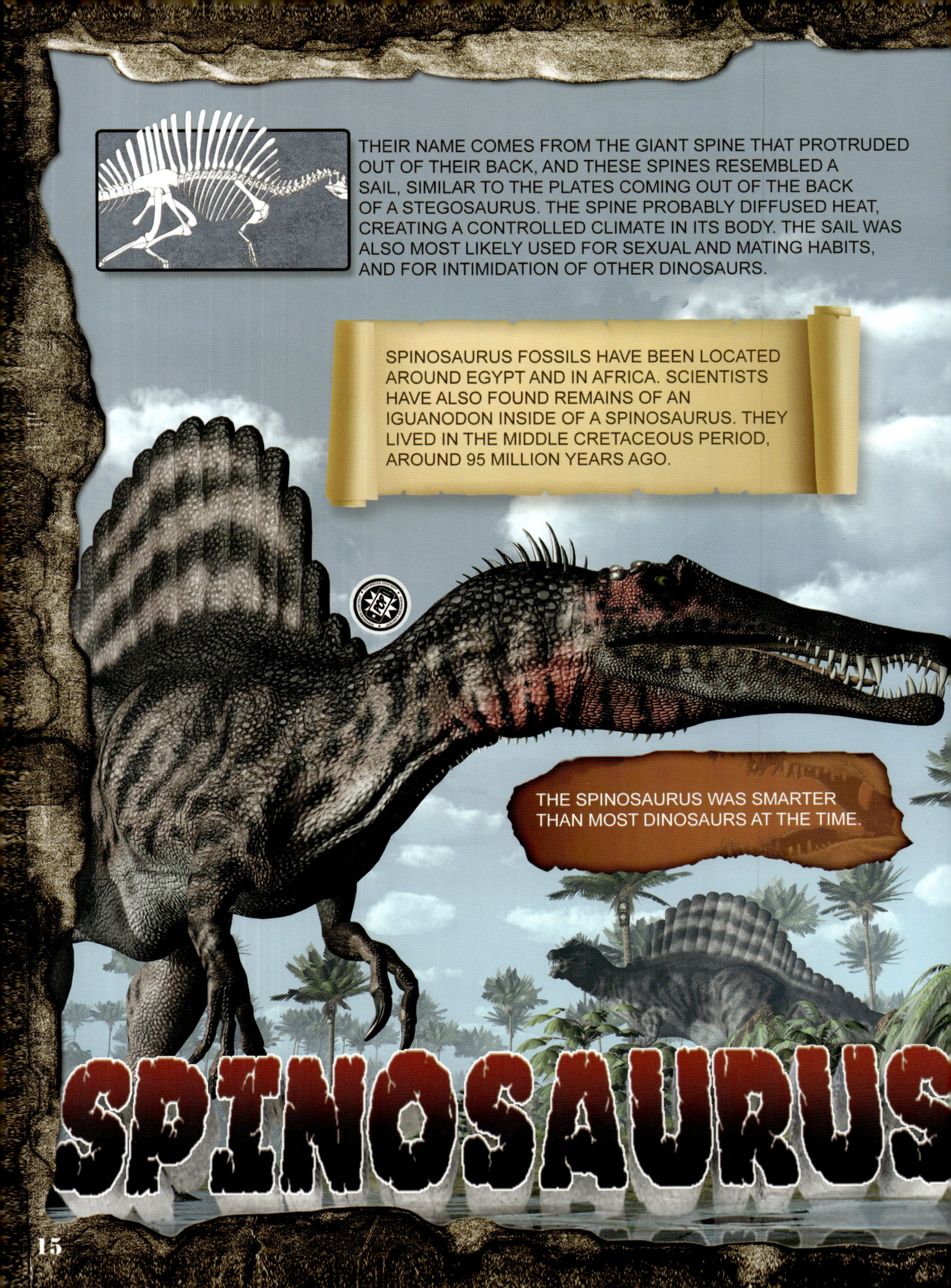

THEIR NAME COMES FROM THE GIANT SPINE THAT PROTRUDED OUT OF THEIR BACK, AND THESE SPINES RESEMBLED A SAIL, SIMILAR TO THE PLATES COMING OUT OF THE BACK OF A STEGOSAURUS. THE SPINE PROBABLY DIFFUSED HEAT, CREATING A CONTROLLED CLIMATE IN ITS BODY. THE SAIL WAS ALSO MOST LIKELY USED FOR SEXUAL AND MATING HABITS, AND FOR INTIMIDATION OF OTHER DINOSAURS.

SPINOSAURUS FOSSILS HAVE BEEN LOCATED AROUND EGYPT AND IN AFRICA. SCIENTISTS HAVE ALSO FOUND REMAINS OF AN IGUANODON INSIDE OF A SPINOSAURUS. THEY LIVED IN THE MIDDLE CRETACEOUS PERIOD, AROUND 95 MILLION YEARS AGO.

THE SPINOSAURUS WAS SMARTER THAN MOST DINOSAURS AT THE TIME.

SPINOSAURUS

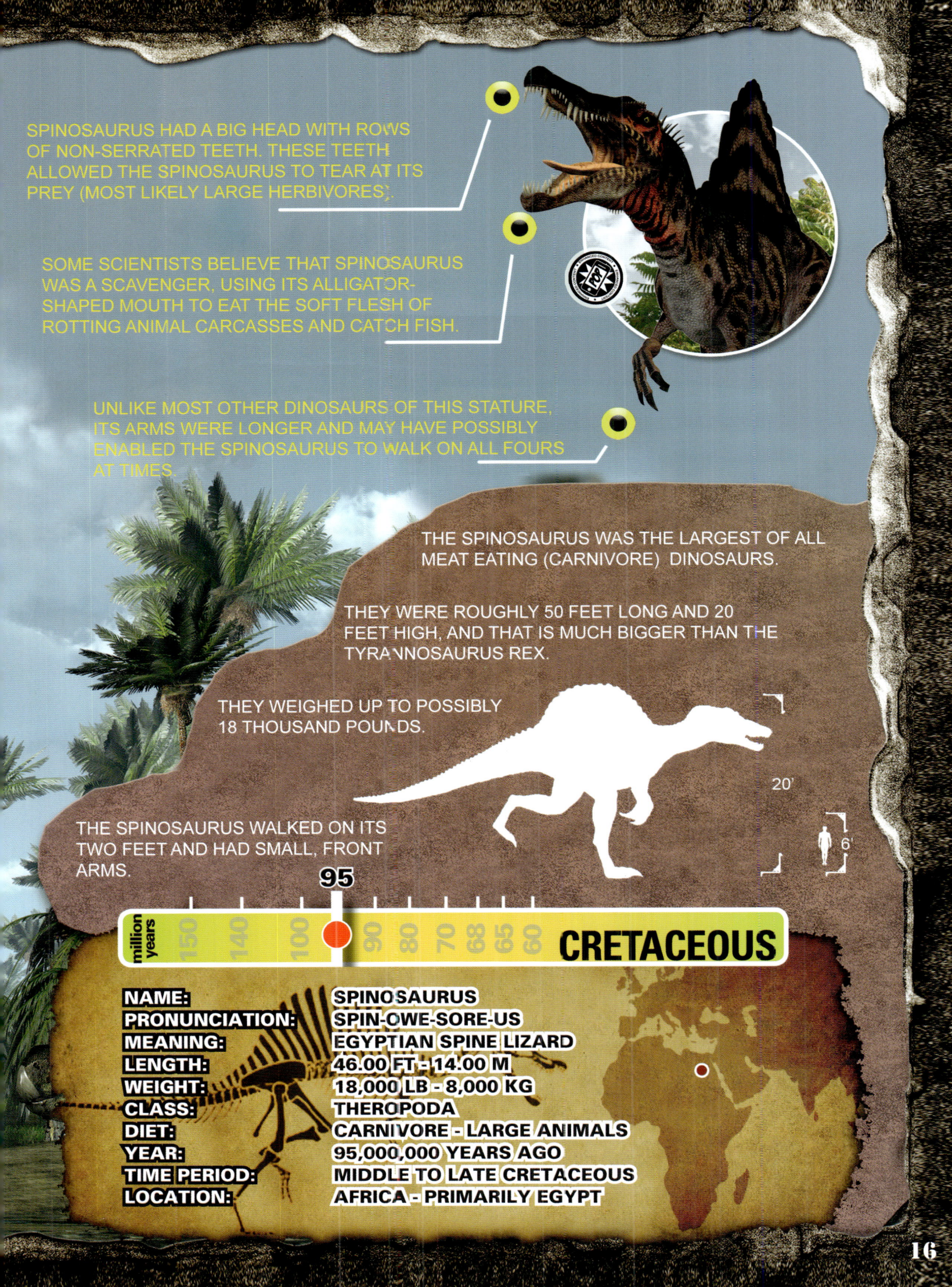

SPINOSAURUS HAD A BIG HEAD WITH ROWS OF NON-SERRATED TEETH. THESE TEETH ALLOWED THE SPINOSAURUS TO TEAR AT ITS PREY (MOST LIKELY LARGE HERBIVORES).

SOME SCIENTISTS BELIEVE THAT SPINOSAURUS WAS A SCAVENGER, USING ITS ALLIGATOR-SHAPED MOUTH TO EAT THE SOFT FLESH OF ROTTING ANIMAL CARCASSES AND CATCH FISH.

UNLIKE MOST OTHER DINOSAURS OF THIS STATURE, ITS ARMS WERE LONGER AND MAY HAVE POSSIBLY ENABLED THE SPINOSAURUS TO WALK ON ALL FOURS AT TIMES.

THE SPINOSAURUS WAS THE LARGEST OF ALL MEAT EATING (CARNIVORE) DINOSAURS.

THEY WERE ROUGHLY 50 FEET LONG AND 20 FEET HIGH, AND THAT IS MUCH BIGGER THAN THE TYRANNOSAURUS REX.

THEY WEIGHED UP TO POSSIBLY 18 THOUSAND POUNDS.

THE SPINOSAURUS WALKED ON ITS TWO FEET AND HAD SMALL, FRONT ARMS.

CRETACEOUS

NAME:	SPINOSAURUS
PRONUNCIATION:	SPIN-OWE-SORE-US
MEANING:	EGYPTIAN SPINE LIZARD
LENGTH:	46.00 FT - 14.00 M
WEIGHT:	18,000 LB - 8,000 KG
CLASS:	THEROPODA
DIET:	CARNIVORE - LARGE ANIMALS
YEAR:	95,000,000 YEARS AGO
TIME PERIOD:	MIDDLE TO LATE CRETACEOUS
LOCATION:	AFRICA - PRIMARILY EGYPT

CRETACEOUS

NAME: VELOCIRAPTOR
PRONUNCIATION: VELL-OSS-EE-RAP-TORE
MEANING: QUICK HUNTER
LENGTH: 5.50 FT - 1.50 M
WEIGHT: 45 LB - 20 KG
CLASS: THEROPODA
DIET: CARNIVORE - SMALL ANIMALS
YEAR: 80,000,000 YEARS AGO
TIME PERIOD: LATE CRETACEOUS
LOCATION: ASIA - PRIMARILY MONGOLIA AND CHINA

THE VELOCIRAPTOR LIVED 80 MILLION YEARS AGO IN THE CRETACEOUS PERIOD. THE VELOCIRAPTOR TENDED TO LIVE IN HOTTER CLIMATES. A DESERT CLIMATE NEAR FRESH WATER IS WHERE MOST RAPTORS WOULD TEND TO LIVE. RAPTOR FOSSILS HAVE BEEN FOUND IN EASTERN ASIA, SPECIFICALLY RUSSIA, CHINA, AND MONGOLIA.

THE VELOCIRAPTOR WAS ONE OF THE SMARTEST ANIMALS IN THE DINOSAUR KINGDOM. THE SMALL NATURE OF THIS ANIMAL FORCED IT TO HUNT IN PACKS MUCH LIKE WOLVES.

RAPTORS USED THEIR SHARP, FOUR-INCH CLAWS TO TEAR AWAY AT ITS PREY.

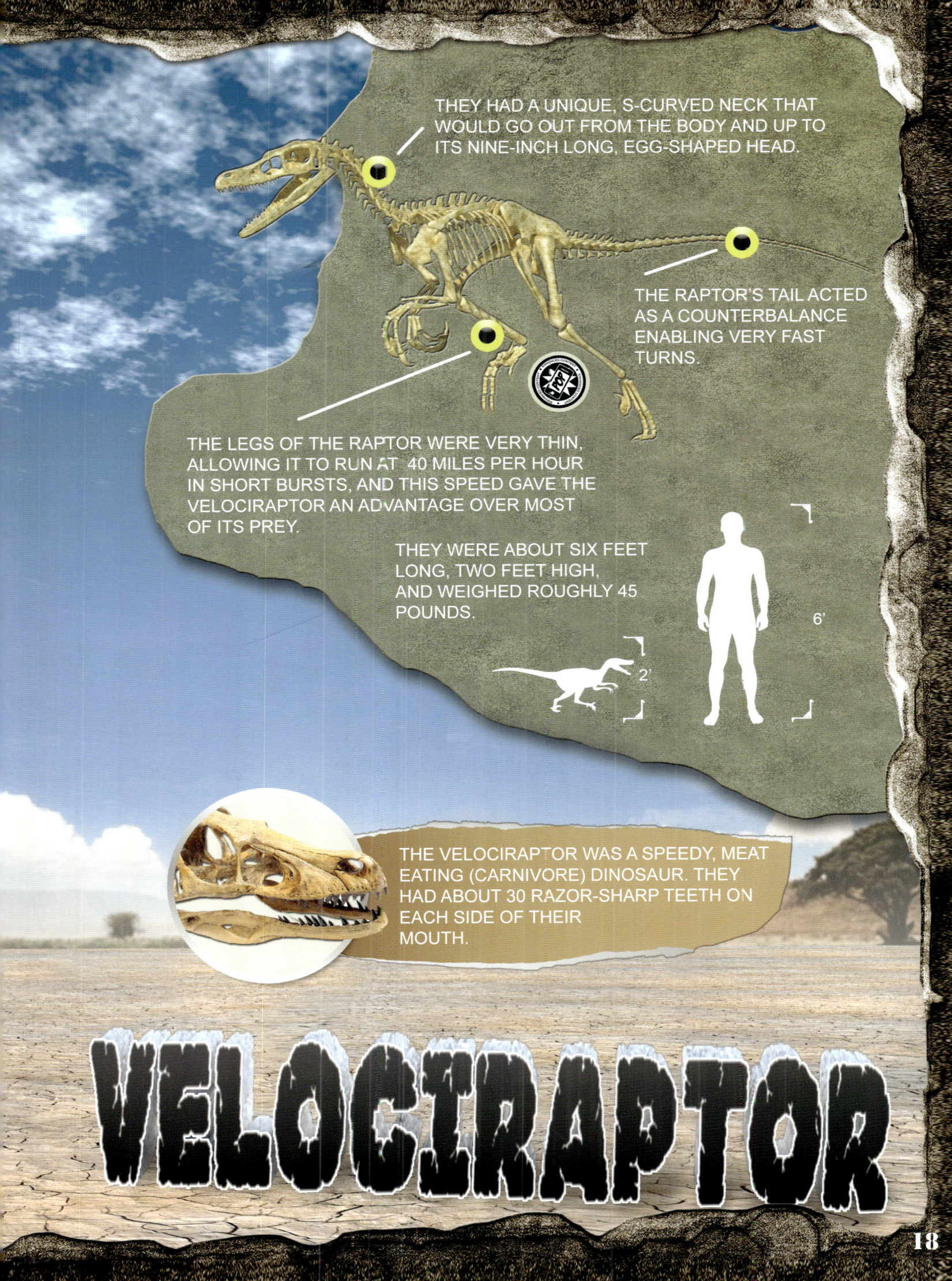

QUETZALCOATLUS

They were one of the only flying reptiles during the period of the dinosaurs.

The Quetzalcoatlus tended to live in caves and trees near the sea. They are thought to have eaten primarily dead animals, fish, and large insects. The leathery material over their long arms caused them to be able to fly quickly.

Quetzalcoatlus lived throughout the Cretaceous and Jurassic time periods. Quetzalcoatlus fossils have been found throughout Europe, North America, Australia, and Africa.

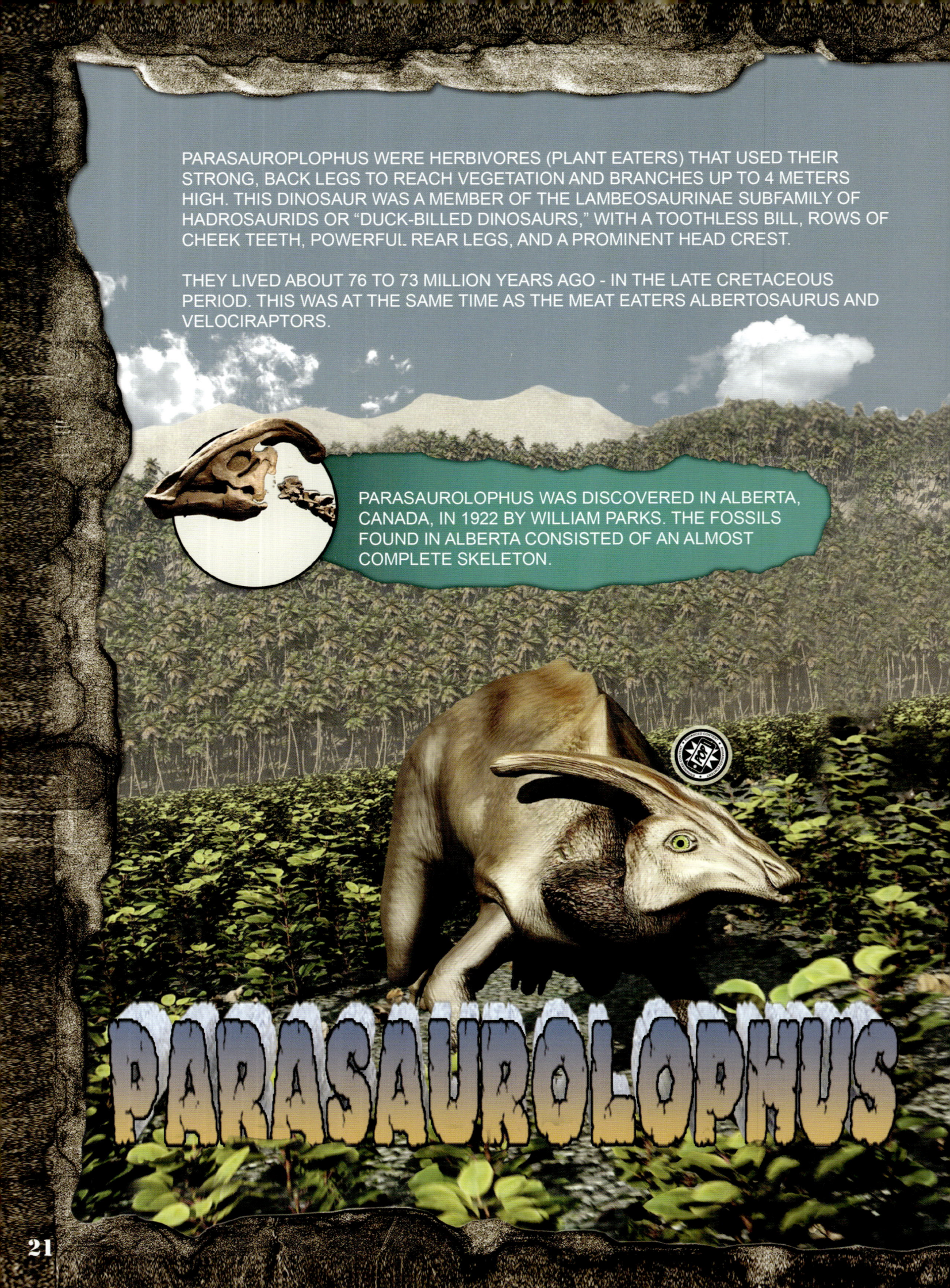

PARASAUROPLOPHUS WERE HERBIVORES (PLANT EATERS) THAT USED THEIR STRONG, BACK LEGS TO REACH VEGETATION AND BRANCHES UP TO 4 METERS HIGH. THIS DINOSAUR WAS A MEMBER OF THE LAMBEOSAURINAE SUBFAMILY OF HADROSAURIDS OR "DUCK-BILLED DINOSAURS," WITH A TOOTHLESS BILL, ROWS OF CHEEK TEETH, POWERFUL REAR LEGS, AND A PROMINENT HEAD CREST.

THEY LIVED ABOUT 76 TO 73 MILLION YEARS AGO - IN THE LATE CRETACEOUS PERIOD. THIS WAS AT THE SAME TIME AS THE MEAT EATERS ALBERTOSAURUS AND VELOCIRAPTORS.

PARASAUROLOPHUS WAS DISCOVERED IN ALBERTA, CANADA, IN 1922 BY WILLIAM PARKS. THE FOSSILS FOUND IN ALBERTA CONSISTED OF AN ALMOST COMPLETE SKELETON.

PARASAUROLOPHUS

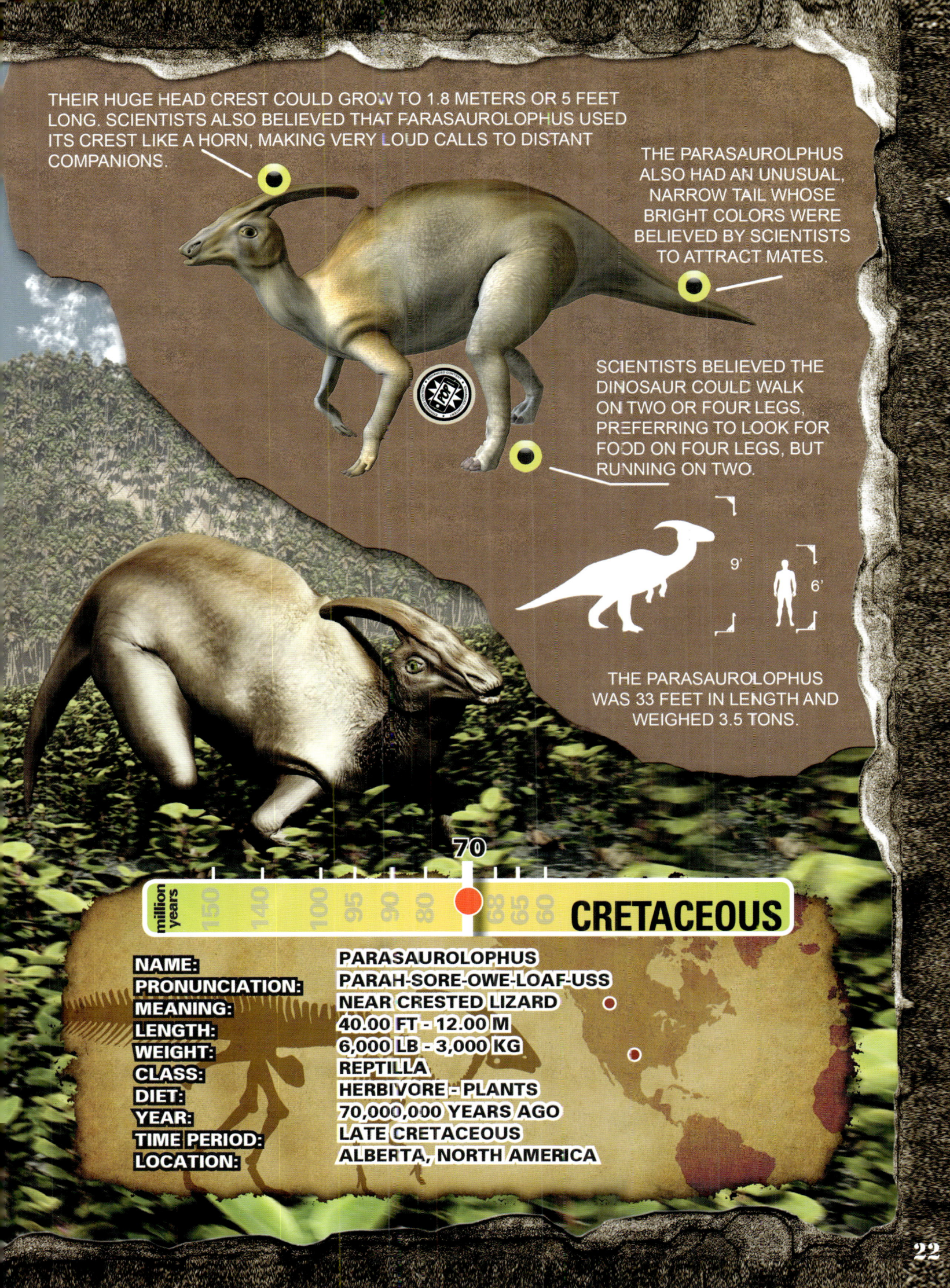

JURASSIC

NAME:	COMPSOGNATHUS
PRONUNCIATION:	COMP-SON-NAY-THUSS
MEANING:	DAINTY JAW
LENGTH:	3.0 FT - 1.0 M
WEIGHT:	1.8-7.7 LB - .83-3.5 KG
CLASS:	THEROPODA
DIET:	CARNIVORE - SMALL ANIMALS
YEAR:	150,000,000 YEARS AGO
TIME PERIOD:	LATE JURASSIC
LOCATION:	EUROPE - PRIMARILY FRANCE AND GERMANY

COMPSOGNATHUS, LIKE MANY OTHER DINOSAURS, HAD HOLLOW BONES MAKING IT VERY LIGHT WEIGHT. THIS IS ONE REASON MANY SCIENTISTS BELIEVE THAT THE BIRDS EVOLVED FROM DINOSAURS LIKE COMPSOGNATHUS.

6'
10"

THE LARGEST SPECIMEN IS ESTIMATED TO HAVE WEIGHED SOMEWHERE BETWEEN 1.8 AND 7.7 POUNDS.

COMPSOGNATHUS HAD A UNIQUELY DESIGNED SKULL. INSTEAD OF SEVERAL PLATES OF BONE FORMING A SKULL AS IN OTHER DINOSAURS, ITS SKULL WAS MADE OF SMALL BITS OF BONE HELD TOGETHER BY FLESH.

COMPSOGNATHUS' USED ITS LONG, SLENDER LEGS AND ELONGATED FEET TO DART QUICKLY ABOUT FROM SIDE TO SIDE AS IT CHASED ITS PREY AND ESCAPED FROM LARGER PREDATORS.

COMPSOGNATHUS IS ONE OF THE FEW DINOSAUR SPECIES WHERE ITS DIET IS KNOWN WITH CERTAINTY. PALEONTOLOGISTS HAVE FOUND TWO WELL-PRESERVED FOSSILS, ONE IN GERMANY IN THE 1850'S AND THE SECOND IN FRANCE MORE THAN A CENTURY LATER. THE REMAINS IN THE GERMAN SPECIMEN SHOW THAT COMPSOGNATHUS PREYED ON SMALL VERTEBRATES. THERE WERE THE REMAINS OF SMALL, AGILE LIZARDS PRESERVED IN THE BELLIES OF BOTH SPECIMENS. THEIR TEETH WERE SMALL BUT SHARP, WELL SUITED FOR ITS DIET OF SMALL VERTEBRATES AND POSSIBLY INSECTS.

COMPSOGNATHUS IS A VERY FAMOUS CREATURE BECAUSE IT IS ONE OF THE SMALLEST KNOWN DINOSAURS WHERE A COMPLETE SKELETON WAS FOUND.

THEY LIVED AROUND 150 MILLION YEARS AGO, THE LATE JURASSIC PERIOD, IN WHAT IS NOW EUROPE.

COMPSOGNATHUS

Digital Toys Here

1 Cut out Popar® paddles on pages 26 and 28

2 Play with Popar® paddles

NEED HELP? GO TO www.PoparToys.com FOR TECH SUPPORT.

STEGOSAURUS

TRICERATOPS

ALLOSAURUS

DIPLODOCUS

T-REX

SPINOSAURUS

Cut out or photocopy the Popar® paddles on this page. Use these paddles with a mobile/smart device and the FREE Popar® app when you are not reading the book. The Popar® paddles have the same amazing 3D objects, just like in the book. These 3D digital toys are great to play with and to take photos to share with your friends and family!

 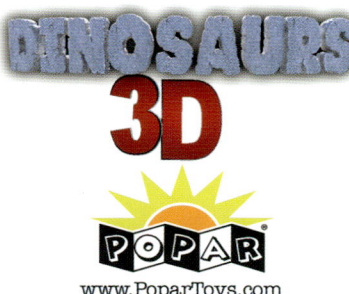 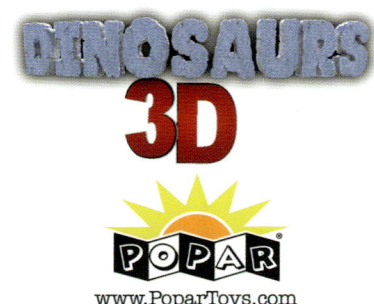

 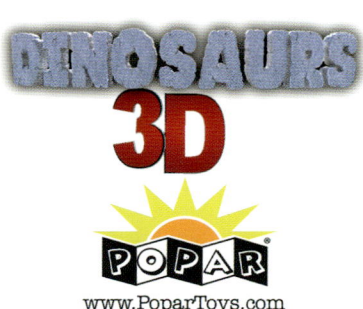 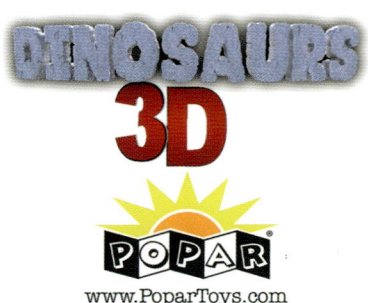

VELOCIRAPTOR **QUETZALCOATLUS** **PARASAUROLOPHUS**

COMPSOGNATHUS

Cut out or photocopy the Popar® paddles on this page. Use these paddles with a mobile/smart device and the FREE Popar® app when you are not reading the book. The Popar® paddles have the same amazing 3D objects, just like in the book. These 3D digital toys are great to play with and to take photos to share with your friends and family!

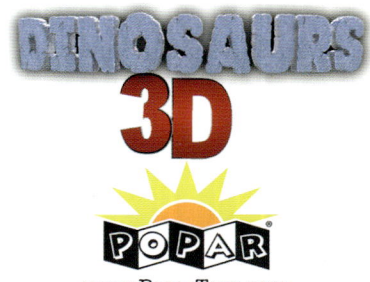 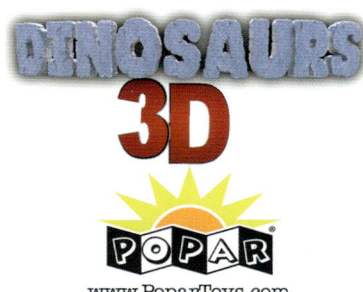 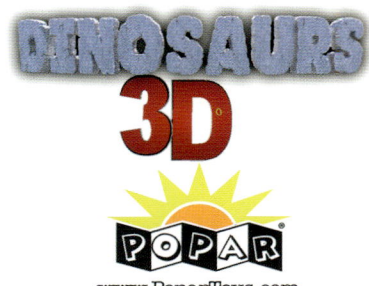